Grade 6 · Unit 2

Inspire Science

Reproduction of Organisms

Mc
Graw
Hill
Education

Grade 6 · Unit 2   Student Edition

Inspire Science
Reproduction of Organisms

Phenomenon: How did these birds build this nest?

The weaver bird, of the family *Ploceidae*, is a native of Africa and tropical areas of Asia. Its name comes from the intricately woven nests that these birds build for protection.

Fun Fact

Evidence has proven that some weaver bird nests are over 100 years old!

mheducation.com/prek-12

Send all inquiries to:
McGraw-Hill Education
STEM Learning Solutions Center
8787 Orion Place
Columbus, OH 43240

ISBN: 978-0-07-687333-3
MHID: 0-07-687333-1

Printed in the United States of America.

4 5 6 7 8 9 QSX 23 22 21 20 19

McGraw-Hill is committed to providing instructional materials in Science, Technology, Engineering, and Mathematics (STEM) that give all students a solid foundation, one that prepares them for college and careers in the 21st century.

Welcome to

Inspire Science

Explore Our Phenomenal World

Learning begins with curiosity. Inspire Science is designed to spark your interest and empower you to ask more questions, think more critically, and generate innovative ideas.

Start exploring now!

Inspire Curiosity • **Inspire Investigation** • **Inspire Innovation**

Program Authors

Alton L. Biggs
Biggs Educational Consulting
Commerce, TX

Ralph M. Feather, Jr., PhD
Professor of Educational Studies and
Secondary Education
Bloomsburg University
Bloomsburg, PA

Douglas Fisher, PhD
Professor of Teacher Education
San Diego State University
San Diego, CA

Page Keeley, MEd
Author, Consultant, Inventor of
Page Keeley Science Probes
Maine Mathematics and Science
Alliance
Augusta, ME

Michael Manga, PhD
Professor
University of California, Berkeley
Berkeley, CA

Edward P. Ortleb
Science/Safety Consultant
St. Louis, MO

Dinah Zike, MEd
Author, Consultant, Inventor
of Foldables®
Dinah Zike Academy, Dinah-Might
Adventures, LP
San Antonio, TX

Advisors

Phil Lafontaine
NGSS Education Consultant
Folsom, CA

Donna Markey
NBCT, Vista Unified School District
Vista, CA

Julie Olson
NGSS Consultant
Mitchell Senior High/Second Chance
High School
Mitchell, SD

Content Consultants

Chris Anderson
STEM Coach and Engineering
Consultant
Cinnaminson, NJ

Emily Miller
EL Consultant
Madison, WI

Key Partners

American Museum of Natural History

The American Museum of Natural History is one of the world's preeminent scientific and cultural institutions. Founded in 1869, the Museum has advanced its global mission to discover, interpret, and disseminate information about human cultures, the natural world, and the universe through a wide-ranging program of scientific research, education, and exhibition.

SpongeLab Interactives

SpongeLab Interactives is a learning technology company that inspires learning and engagement by creating gamified environments that encourage students to interact with digital learning experiences. Students participate in inquiry activities and problem-solving to explore a variety of topics through the use of games, interactives, and video while teachers take advantage of formative, summative, or performance-based assessment information that is gathered through the learning management system.

PhET Interactive Simulations

The PhET Interactive Simulations project at the University of Colorado Boulder provides teachers and students with interactive science and math simulations. Based on extensive education research, PhET simulations engage students through an intuitive, game-like environment where students learn through exploration and discovery.

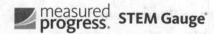

Measured Progress, a not-for-profit organization, is a pioneer in authentic, standards-based assessments. Included with New York Inspire Science is **Measured Progress STEM Gauge®** assessment content which enables teacher to monitor progress toward learning NGSS.

Table of Contents
Reproduction of Organisms

Reproduction of Organisms

How do these Kokanee salmon reproduce and grow?

Spawning
Salmon

GO ONLINE
Watch the video *Spawning Salmon* to see this phenomenon in action.

Communicate Kokanee salmon return to the same streams in which they were born to mate. Record your ideas for why you think this happens. Discuss your ideas with three different partners. Revise or update your ideas, if necessary, after the discussions with your classmates.

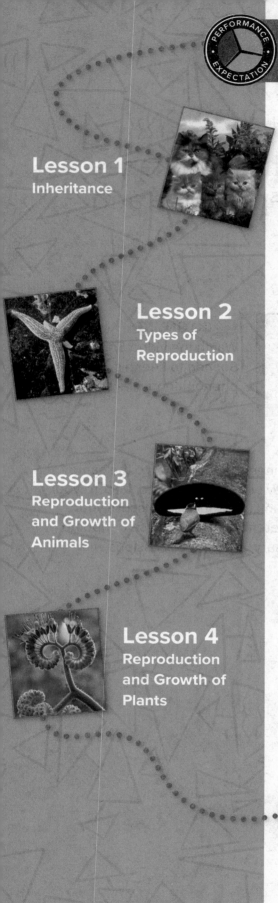

Lesson 1
Inheritance

Lesson 2
Types of
Reproduction

Lesson 3
Reproduction
and Growth of
Animals

Lesson 4
Reproduction
and Growth of
Plants

Get Your Game Face On

A scientific toy company has asked your team to develop and test a game that models the differences in offspring produced through asexual and sexual reproduction. The goals of the game are for players to ensure the successful reproduction of an alien species that can reproduce through different methods.

Your game will include behaviors as well as specialized structures that affect the probability of reproduction, environmental and genetic factors influencing the growth of organisms, and senses that help with reproduction and survival, such as those that help Kokanee salmon return to their home streams to spawn.

Start Thinking About It

In the image above you see a puppy and its mother. Why does the puppy look different from its mother? Do all offspring look different from their parents? Discuss your thoughts with your group.

STEM Module Project
Planning and Completing the Science Challenge
How will you meet this goal? The concepts you will learn throughout this module will help you plan and complete the Science Challenge. Just follow the prompts at the end of each lesson!

PAGE KEELEY
SCIENCE
PROBES

Ladybugs

Alice and her friends looked at a bunch of ladybugs. They noticed the ladybugs had different numbers of spots. They each had different ideas about why the numbers of spots were different. This is what they said:

Fred: I think it has to do with the sex of the ladybug. Males will add more spots to compete for the females.

Andrea: I think it depends on what each ladybug was born with.

Karen: I think they must all be different species of ladybugs.

Isaac: I think it depends on the age of the ladybugs. As they get older they add more spots.

Alice: I think they change their number of spots when there is a need for them to do so.

Troy: I think each ladybug has the same number of spots as its parents.

Which person do you agree with the most? Explain your thinking. Describe your ideas about why the ladybugs look different.

_____

_____

_____

_____

You will revisit your response to the Science Probe at the end of the lesson.

Inheritance

ENCOUNTER
THE PHENOMENON

Why do some offspring look like their parents, while others do not?

Observe the images provided by your teacher. Which kittens do you think belong to each group of parents? Record your thoughts below.

Kittens

GO ONLINE
Watch the video *Kittens* to see the phenomenon in action.

EXPLAIN
THE PHENOMENON

Did you see how some of the kittens looked like their parent, but some did not? Use your observations about the phenomenon to make a claim about why offspring sometimes look like their parents.

CLAIM

Offspring sometimes look like their parents...

 COLLECT EVIDENCE as you work through the lesson. Then return to these pages to record your evidence.

EVIDENCE

A. What evidence have you discovered to explain how Mendel's experiments showed that traits in offspring, such as those of the kittens, are inherited?

B. What evidence have you discovered to explain the factors that control traits in offspring, such as those of the kittens?

MORE EVIDENCE

C. What evidence have you discovered to explain how inheritance of traits in offspring, such as kittens, can be modeled?

When you are finished with the lesson, review your evidence. If necessary, based on the evidence, revise your claim.

REVISED CLAIM

Offspring sometimes look like their parents...

Finally, explain your reasoning for how and why your evidence supports your claim.

REASONING

The evidence I collected supports my claim because...

What are traits?

Your characteristics are what make you unique. They could be things like hair color or height. These characteristics are called traits. How a trait appears, or is expressed, is the trait's **phenotype** (FEE nuh tipe). Traits such as eye color have many different types, but some traits have only two types.

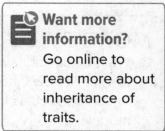

Want more information?
Go online to read more about inheritance of traits.

INVESTIGATION

Understanding Traits

By a show of hands, determine how many students in your class have each type of trait below. Write your observations in your Science Notebook.

Student Traits		
Trait	**Type 1**	**Type 2**
Earlobes	Unattached	Attached
Thumbs	Curved	Straight
Interlacing fingers	Left thumb over right thumb	Right thumb over left thumb

What do you think determines the types of traits you have?

_____

_____

_____

You have just observed a variety of traits within your classroom. These traits were passed to your classmates from their parents. **Heredity,** the passing of traits from parents to offspring, is complex. For example, you might have straight thumbs, but both of your parents have curved thumbs.

Why do offspring look like their parents?

You may have noticed that the kittens at the beginning of the lesson looked similar to their parents. How did the traits of the parents pass to the offspring?

FOLDABLES
Go to the Foldables® library to make a Foldable® that will help you take notes while reading this lesson.

HISTORY ⟩ Connection More than 150 years ago, Gregor Mendel, an Austrian monk, performed experiments that helped answer this question. Because of his research, Mendel is known as the father of **genetics** (juh NEH tihks)—the study of how traits are passed from parents to offspring.

Mendel performed controlled breeding experiments with pea plants. He began with plants that were true-breeding for a certain trait. When a true-breeding plant pollinates itself, it always produces offspring with traits that match the parent. For example, when a true-breeding pea plant with wrinkled seeds self-pollinates, it produces only plants with wrinkled seeds.

Observe some of Mendel's findings from cross-pollinating (one plant pollinates another) true-breeding plants below. What do you think Mendel discovered when he crossed true-breeding plants with purple flowers and true-breeding plants with white flowers? Illustrate your predictions below.

Purple × Purple

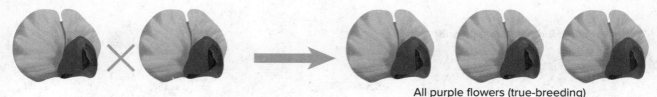

All purple flowers (true-breeding)

White × White

All white flowers (true-breeding)

Purple (true-breeding) × White (true-breeding)

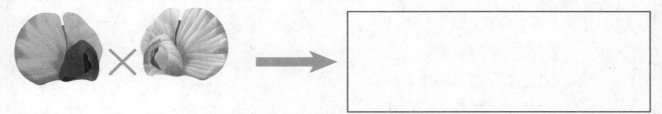

First-Generation Crosses A cross between two true-breeding plants with purple flowers produced plants with only purple flowers. A cross between two true-breeding plants with white flowers produced plants with only white flowers. But something unexpected happened when Mendel crossed true-breeding plants with purple flowers and true-breeding plants with white flowers—all the offspring had purple flowers.

Why didn't the cross produce offspring with light purple flowers—a combination of the white and purple flower colors? Mendel carried out more experiments with pea plants to answer this question.

Second-Generation (Hybrid) Crosses The first-generation purple-flowering plants are called hybrid plants. This means they came from true-breeding parent plants with different forms of the same trait. Mendel wondered what would happen if he cross-pollinated two purple-flowering hybrid plants. Take a look at his second-generation cross results below.

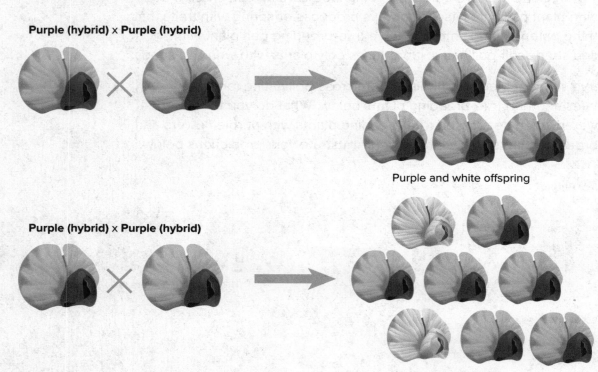

Purple (hybrid) × **Purple (hybrid)**

Purple and white offspring

Purple (hybrid) × **Purple (hybrid)**

Purple and white offspring

THREE-DIMENSIONAL THINKING

What patterns do you notice in the results of Mendel's second-generation cross between hybrid plants with purple flowers? Explain how the results may have occurred.

_____

_____

_____

Reappearing Traits As you observed, some of the offspring had white flowers, even though both parents had purple flowers. The results were similar each time Mendel cross-pollinated two purple-flowering hybrid plants. The trait that had disappeared in the first generation always reappeared in the second generation.

Did the same result happen when Mendel cross-pollinated pea plants for other traits? Let's find out.

INVESTIGATION

Look Both Ways Before Crossing the Seed

Mendel counted and recorded the traits of offspring from many experiments in which he cross-pollinated hybrid plants. Data from these experiments are shown below.

Results of Hybrid Crosses			
Characteristic of Hybrid Parent	Trait and Number of Offspring	Trait and Number of Offspring	Trait Comparison
Flower Color (purple x purple)	Purple 705	White 224	$\frac{705}{224} = \frac{}{1}$
Seed Color (yellow x yellow)	Yellow 6,022	Green 2,001	$\frac{6,022}{2,001} = \frac{}{1}$
Seed Shape (round x round)	Round 5,474	Wrinkled 1,850	$\frac{5,474}{1,850} = \frac{}{1}$
Pod Shape (smooth x smooth)	Smooth 882	Bumpy 299	$\frac{882}{299} = \frac{}{1}$

1. **MATH Connection** Calculate the relationship of purple to white flowers, yellow to green seeds, round to wrinkled seeds, and smooth to bumpy pods by dividing the higher number by the lower number. Record the answers in the table above.

2. What patterns do you notice in Mendel's data?

A Similar Relationship From the data of crosses between hybrid plants with purple flowers, Mendel found that the relationship of purple flowers to white flowers was about 3 to 1. This means purple-flowering pea plants grew from this cross three times more often than white-flowering pea plants grew from the cross. He calculated similar relationships for all traits he tested.

Dominant and Recessive Traits Recall that when Mendel cross-pollinated a true-breeding plant with purple flowers and another with white flowers, the hybrid offspring had only purple flowers. Mendel hypothesized that the hybrid offspring had one genetic factor for purple flowers and one genetic factor for white flowers.

Mendel also hypothesized that the purple factor is the only factor seen or expressed because it blocks the white factor. A genetic factor that blocks another genetic factor is called a **dominant** (DAH muh nunt) trait. A genetic factor that is blocked by the presence of a dominant factor is called a **recessive** (rih SE sihv) trait.

A dominant trait, such as purple pea flowers, is observed when offspring have either one or two dominant factors.

A recessive trait, such as white pea flowers, is observed only when two recessive genetic factors are present in offspring.

COLLECT EVIDENCE

What did Mendel's experiments show about how the traits of offspring, such as those of the kittens, are inherited? Record your evidence (A) in the chart at the beginning of the lesson.

THREE-DIMENSIONAL THINKING

Now that you have learned about dominant and recessive traits, take a look back at the table on the previous page. **Construct an explanation** for which seed color is the dominant trait.

What controls traits?

When other scientists studied the parts of a cell and combined Mendel's work with their work, Mendel's factors were more clearly understood. Scientists discovered that inside each cell is a nucleus that contains threadlike structures called chromosomes. Over time, scientists learned that chromosomes contain genetic information that controls traits. We now know that Mendel's "factors" are parts of chromosomes and that each cell in offspring contains chromosomes from both parents. These chromosomes exist as pairs—one chromosome from each parent.

Scientists have discovered that each chromosome can have information about hundreds or even thousands of traits.

- A **gene** (JEEN) is a section on a chromosome that has genetic information for one trait. The genes on each chromosome can be the same or different, such as purple or white for pea flower color.
- The different forms of a gene are called **alleles** (uh LEELs).
- The two alleles that control the phenotype of a trait are called the trait's **genotype.**

Scientists use symbols to represent the alleles in a genotype, as shown in the table below. In genetics, uppercase letters represent dominant alleles and lowercase letters represent recessive alleles. The table shows the possible genotypes for both round and wrinkled seeds phenotypes.

Phenotype and Genotype			
Phenotypes (observed traits)	 **Round**		 **Wrinkled**
Genotypes (alleles of a gene)	Homozygous dominant (*RR*)	Heterozygous (*Rr*)	Homozygous recessive (*rr*)

A round seed can have two genotypes—*RR* and *Rr*. Both genotypes have a round phenotype. A wrinkled seed can have only one genotype—*rr*.

- When the two alleles of a gene are the same, its genotype is **homozygous.**
- Both *RR* and *rr* are homozygous genotypes.
- If the two alleles of a gene are different, its genotype is **heterozygous.**
- *Rr* is a heterozygous genotype.

LAB  Beetle Genes

You have learned that two alleles make up a trait's genotype. If you know that beetle traits are either dominant or recessive, how can you use this information to model the genotype of a beetle?

Materials

trait bags (3)
colored pencils

Procedure

1. Select one trait card from each of the three beetle trait bags. Record the traits you selected in the *Individuals with Trait* column in the table below.

2. Draw a picture of your beetle based on the traits you selected.

3. Combine your data with the rest of the class. Record the total number of individuals with each trait.

Beetle Traits				
Phenotype	Individuals with Trait	Dominant	Recessive	Genotypes
Green body				
Red body				
Round Spots				
No Spots				
Long wings				
Short wings				

Analyze and Conclude

4. Describe any patterns you find in the data table.

_____

_____

_____

5. Determine which trait is dominant and which is recessive. Record your responses in the table. Explain your reasoning.

_____

_____

_____

_____

6. Determine the possible genotype(s) for each phenotype. Record your responses in the table. Explain your reasoning.

_____

_____

_____

_____

7. Decide whether you could have correctly determined your beetle's genotype without data from other beetles. Explain your reasoning.

_____

_____

_____

_____

COLLECT EVIDENCE

What factors control traits, such as those of the kittens at the beginning of the lesson? Record your evidence (B) in the chart at the beginning of the lesson.

How can you predict what offspring will look like?

Have you ever flipped a coin and guessed heads or tails? Because a coin has two sides, there are only two possible outcomes—heads or tails. You have a 50 percent chance of getting heads and a 50 percent chance of getting tails. The ratio comparing 50 heads to 50 tails can be written 50 to 50 or 50 : 50, or simplified, 1 : 1.

MATH ▶Connection A ratio is a comparison of two numbers or quantities by division. For example, the ratio comparing **6,022** yellow seeds to **2,001** green seeds can be written as follows:

$$6,022 \text{ to } 2,001 \text{ or}$$

$$6,022 : 2,001 \text{ or}$$

$$\frac{6,022}{2,001}$$

To simplify the ratio, divide the first number by the second number.

$$\frac{6,022}{2,001} = \frac{3}{1} \text{ or } 3:1$$

Given a 3 : 1 ratio, you can expect that an offspring from heterozygous parents has a 3 : 1 chance of having yellow seeds. But you cannot expect that a group of four seeds will have three yellow seeds and one green seed. This is because one offspring does not affect the phenotype of another offspring. In a similar way, the outcome of one coin toss does not affect the outcome of other coin tosses.

 THREE-DIMENSIONAL THINKING

A cross between two heterozygous pea plants with yellow seeds produced 1,719 yellow seeds and 573 green seeds. What is the ratio of yellow to green seeds? **Construct an explanation** about what the results show regarding inheritance.

_____

_____

_____

_____

_____

_____

Punnett Squares If the genotypes of the parents are known, then the different genotypes and phenotypes of the offspring can be predicted. A Punnett square is a model used to predict possible genotypes and phenotypes of offspring.

Punnett Predictions

Suppose that you are trying to predict the outcome of a cross between two hybrid plants with yellow pea seeds (genotypes *Yy*). Recall that yellow seeds are dominant, while green seeds are recessive. To determine the phenotypes of the offspring, follow the prompts and fill in the figure below.

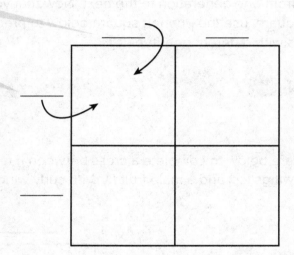

1 Split the alleles from the male genotype, *Yy*, and place one letter on each column.

2 Split the alleles from the female genotype, *Yy*, and place one letter on each row.

3 Copy female alleles across each row. Copy male alleles down each column. Always list the dominant trait first. The arrows show how to fill in the first square.

4 Add in which genotype will display yellow peas and which will display green peas.

So, how did you do? Your completed Punnett square should look like this:

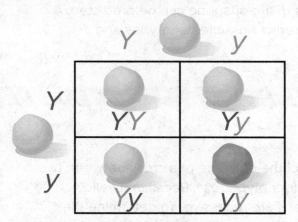

Geneticists, or scientists who study genetics, use Punnett squares to explain how traits are inherited from one generation to the next. Now that you know how to fill in a Punnett square, use the Punnett square below to predict the offspring of two fruit flies with different genotypes.

INVESTIGATION

Fruit Fly Traits

1. Use the Punnett square below to complete a cross between a female fruit fly with straight wings (cc) and a male fruit fly with curly wings (CC).

Curly wings (CC)

Straight wings (cc)

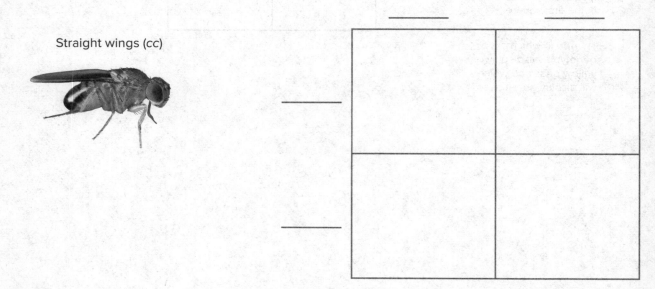

2. According to your Punnett square, which genotypes are possible in the offspring?

_____

3. If you switch the locations of the parent genotypes around the Punnett square, does it affect the potential genotypes of their offspring? Explain.

_____

_____

_____

4. Based on the information in your Punnett square, what is the ratio of offspring that will have curly wings to straight wings?

_____

_____

_____

THREE-DIMENSIONAL THINKING

Design and complete a Punnett square to **model** a cross between two fruit flies that are heterozygous for curly wings (*Cc*). What is the phenotype ratio of the offspring? **Explain** how you are able to predict the phenotype ratio based on a **cause and effect** relationship.

How can you model a family's phenotypes?

Another model that can show inherited traits is a pedigree. A **pedigree** shows phenotypes of genetically related family members. It can also help determine genotypes. In the pedigree below, three offspring have a trait— attached earlobes—that the parents do not have. If these offspring received one allele for this trait from each parent, but neither parent displays the trait, the offspring must have received two recessive alleles.

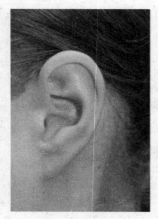

Attached lobe

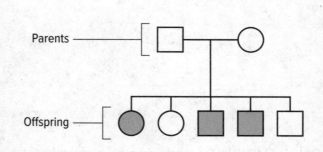

Parents

Offspring

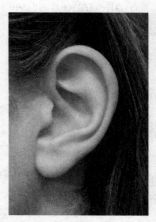

Unattached lobe

Recessive phenotype	Dominant phenotype
⬤ Female with attached lobes	◯ Female with unattached lobes
⬛ Male with attached lobes	☐ Male with unattached lobes

THREE-DIMENSIONAL THINKING

If the genotype of the offspring with attached lobes is *uu*, what is the genotype of the parents? **Explain** your answer.

COLLECT EVIDENCE

How can inheritance of traits in offspring, such as kittens, be modeled? Record your evidence (C) in the chart at the beginning of the lesson.

A Day in the Life of a Genetic Counselor

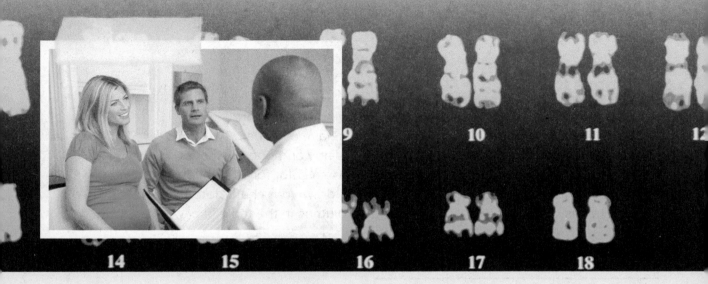

Genetic counselors identify specific genetic disorders or risks through the study of genetics. Most genetic disorders or syndromes are inherited. For parents who are expecting children, counselors use genetics to predict whether a baby is likely to have hereditary disorders, such as Down syndrome and cystic fibrosis, among others. Genetic counselors also assess the risk for an adult to develop diseases with a genetic component, such as certain forms of cancer.

Genetic counselors typically interview patients to obtain comprehensive individual family and medical histories. They discuss testing options and the associated risks, benefits, and limitations with patients and families as well. They counsel patients and family members by providing information, education, or reassurance regarding genetic risks and inherited conditions.

Counselors identify these conditions by studying patients' genes through DNA testing. Medical laboratory technologists perform lab tests, which genetic counselors then evaluate and use for counseling patients and their families. They then share this information with other health professionals.

It's Your Turn

READING Connection Imagine that you are a genetic counselor tasked with presenting research findings to your colleagues. Choose a genetic disorder to research and present. Include the causes and possible treatments. Integrate visual displays to clarify your findings.

Summarize It!

1. **Model** a pedigree chart that reflects the following information: Two parents have five children. Both of the parents have curly hair. Two boys and one girl have curly hair; the other two, one boy and one girl, have straight hair. Before you draw your chart, choose a color for straight and curly hair, and indicate it in the table. After you draw your chart, determine which trait is dominant and label the proper columns in the table.

_____ phenotype	_____ phenotype
◯ Female with curly hair	⬤ Female with straight hair
▢ Male with curly hair	▮ Male with straight hair

Three-Dimensional Thinking

Susana visits with four generations of her family. Her great-grandmother shows her an old family photo of Susana's great-aunts and great-uncles when they were children. Susana is surprised to see that one of the great-uncles looks almost exactly like her younger brother does now. They have the same distinctive hairline and eye shape. Her great-grandmother tells her that it is the result of heredity.

2. Which is the best explanation that shows the sequence of inheritance that led to Susana having a brother who has the same hairline and eye shape as her great-uncle?

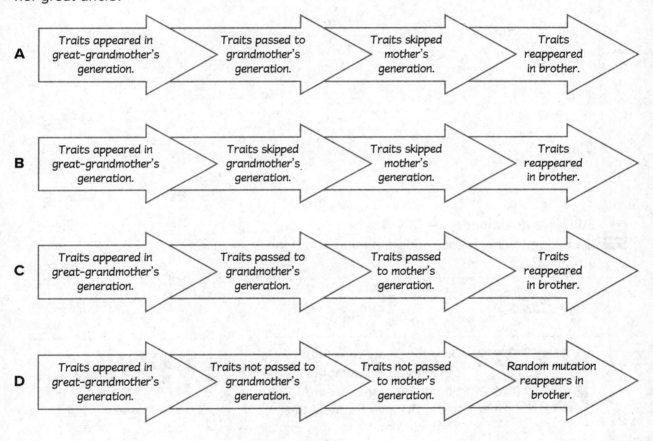

A   Traits appeared in great-grandmother's generation. → Traits passed to grandmother's generation. → Traits skipped mother's generation. → Traits reappeared in brother.

B   Traits appeared in great-grandmother's generation. → Traits skipped grandmother's generation. → Traits skipped mother's generation. → Traits reappeared in brother.

C   Traits appeared in great-grandmother's generation. → Traits passed to grandmother's generation. → Traits passed to mother's generation. → Traits reappeared in brother.

D   Traits appeared in great-grandmother's generation. → Traits not passed to grandmother's generation. → Traits not passed to mother's generation. → Random mutation reappears in brother.

3. When Mendel crossed a true-breeding plant with purple flowers and a true-breeding plant with white flowers, ALL offspring had purple flowers. The best explanation for this data is that the white flowers are

   A dominant.

   B heterozygous.

   C recessive.

   D neutral.

Real-World Connection

4. **Infer** Why have Mendel's experiments been so valuable?

_____

_____

_____

_____

5. **Construct an explanation** for why someone who grows plants might find a Punnett square to be a useful tool.

_____

_____

_____

_____

_____

_____

_____

 Still have questions?
Go online to check your understanding about the inheritance of traits.

 **REVISIT** PAGE KEELEY **SCIENCE PROBES**

Do you still agree with the person you chose at the beginning of the lesson? Return to the Science Probe at the beginning of the lesson. Explain why you agree or disagree with that person now.

START PLANNING STEM Module Project Science Challenge

Now that you've learned why offspring sometimes look like their parents, go back to your Module Project to start planning your game. You will want to model how traits, such as those of the Kokanee salmon, are inherited.

 **EXPLAIN** THE PHENOMENON

Revisit your claim about why offspring sometimes look like their parents. Review the evidence you collected Explain how your evidence supports your claim.

Do they produce eggs?

Five friends looked at a chicken egg. They wondered if other organisms, besides chickens, produce eggs for reproduction. They each had a different idea.

Eddie: I think only birds produce eggs.

Vernon: I think only amphibians, reptiles, and birds produce eggs.

Forest: I think all animals, except mammals, produce eggs.

Angie: I think all animals, except humans, produce eggs.

Sophie: I think all animals, including humans, produce eggs.

Who do you most agree with and why? Describe your ideas about eggs and reproduction.

_____

_____

_____

_____

_____

_____

You will revisit your response to the Science Probe at the end of the lesson.

Types of Reproduction

ENCOUNTER
THE PHENOMENON | How does this sea star reproduce?

Sea stars may be interesting to watch, but they can also be pests to fishermen. At one time, fishermen noticed that sea stars were eating all the clams and other mollusks they were trying to catch. The fishermen tried to get rid of the sea stars by chopping them up and throwing them back into the ocean. When the fishermen returned to the area weeks later, they noticed that the number of sea stars had grown much larger. How is this possible? Make a diagram illustrating how you think the sea star population grew in size.

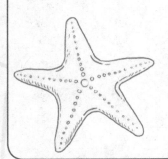

Sea-ing Stars

GO ONLINE
Watch the animation *Sea-ing Stars* to see this phenomenon in action.

EXPLAIN
THE PHENOMENON

You observed a photo of a sea star and hypothesized why more sea stars appeared even after fisherman chopped them up and threw them back in the ocean. Are you starting to get some ideas about how different organisms reproduce? Use your ideas to make a claim about how you think the sea stars reproduced.

CLAIM

More sea stars appeared...

COLLECT EVIDENCE as you work through the lesson.

Then return to these pages to record your evidence.

EVIDENCE

A. What evidence have you discovered to explain how one organism such as the sea star produces offspring that are identical to itself?

B. What evidence have you discovered to explain how two organisms produce offspring that are not identical to themselves?

MORE EVIDENCE

C. What are the advantages and disadvantages of sexual and asexual reproduction?

When you are finished with the lesson, review your evidence. If necessary, based on the evidence, revise your claim.

REVISED CLAIM

More sea stars appeared...

Finally, explain your reasoning for how and why your evidence supports your claim.

REASONING

The evidence I collected supports my claim because...

How can one organism make more organisms?

You learned about many sea stars appearing when they were thought to have been removed from the environment. How is it possible that the sea stars increased their numbers after they had been chopped up?

INVESTIGATION

Plant Progeny

Observe two plants—a seed potato and a coleus stem—in glasses of water. Look at photos of the plants when they were first placed in water. Draw a detailed diagram of each of the glasses in your Science Notebook. Observe the plants a week after placement in the water and write down your observations in your Science Notebook.

1. How did the potato and the coleus plant change after one week?

_____

_____

2. How do you think that this relates to the sea stars you heard about in the introduction to this lesson?

_____

_____

Asexual Reproduction Recall in the beginning of the lesson you read about sea stars reproducing after being cut up. This happened through regeneration. **Regeneration** occurs when an offspring grows from a piece of its parent. Just like sea stars, some plants don't need seeds to make new plants. Some plants can be grown from a leaf, a stem, or another plant part through vegetative reproduction. **Vegetative reproduction** is a form of asexual reproduction in which offspring grow from a part of a parent plant. **Asexual reproduction** occurs when only one parent organism or part of that organism produces a new organism. The new organism is genetically identical to the parent.

> **Want more information?**
> Go online to read more about the different types of reproduction.

> **FOLDABLES**
> Go to the Foldables® library to make a Foldable® that will help you take notes while reading this lesson.

How do other organisms reproduce asexually?

You just discovered asexual reproduction in plants. Different types of plants, animals, and other organisms can reproduce asexually. However, not all asexually reproducing organisms follow the same type of reproduction. Take a look at the hydra below to see how it reproduces.

INVESTIGATION

Break Off a Piece

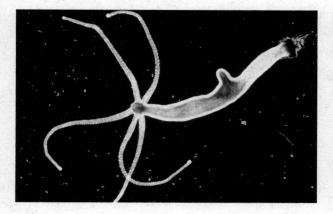 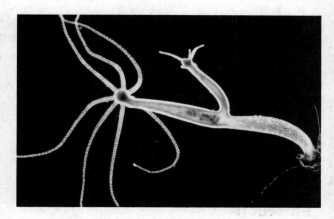

1. Examine the pictures of the hydra above. What evidence do you observe that the hydra reproduced?

_____

_____

_____

2. What are some advantages and disadvantages of this type of reproduction?

_____

_____

_____

The hydra seen in the photos is budding. **Budding** is a form of asexual reproduction where an organism grows on the body of its parent.

COLLECT EVIDENCE
How can one organism such as the sea star produce offspring that are identical to itself? Record your evidence (A) in the chart at the beginning of the lesson.

Why do some organisms have two parents?

When an organism reproduces asexually, its offspring are genetically identical to the parent. Why aren't all organisms identical?

LAB Modeling Offspring

Unless you're an identical twin, you probably don't look exactly like any siblings you might have. You may have differences in physical characteristics such as eye color, hair color, ear shape, or height. Why are there differences in the offspring from the same parents?

Materials

assorted colored beads   markers

paper bags (2)

Procedure

1. Read and complete a lab safety form.

2. Open the paper bag labeled *Male Parent*, and, without looking, remove three beads. Record the bead colors in the Data and Observations section, and replace the beads.

3. Open the paper bag labeled *Female Parent*, and, without looking remove three beads. Record the bead colors, and replace the beads.

4. Repeat steps 2 and 3 for each member of the group. Record your total bead color combinations. Each combination of male and female beads represents an offspring.

5. Follow your teacher's instructions for proper cleanup.

Data and Observations

Analyze and Conclude

6. Compare your group's offspring to another group's offspring. What do you notice?

_____

_____

7. Why do you think there are many different combinations of bead colors?

_____

_____

_____

8. What caused any differences you observed? Explain.

_____

_____

_____

9. Why might this type of reproduction be beneficial to an organism?

_____

_____

_____

You just modeled the genetic result of organisms reproducing by using two parents through sexual reproduction. **Sexual reproduction** is a type of reproduction in which the genetic materials from two different cells combine, producing offspring that are genetically different from their parents. In sexual reproduction, each parent contributes half of the genes aquired by the offspring. Offspring have two of each chromosome and therefore two alleles of each gene, one acquired from each parent.

COLLECT EVIDENCE

How can two organisms produce offspring that are not identical to themselves, unlike the sea star at the beginning of the lesson? Record your evidence (B) in the chart at the beginning of the lesson.

What are the advantages and disadvantages of sexual and asexual reproduction?

	Asexual Reproduction	Sexual Reproduction
Advantages	Asexual reproduction enables organisms to reproduce without a mate. Searching for a mate takes time and energy. Asexual reproduction also enables some organisms to rapidly produce a large number of offspring.	Genetic variation occurs in all organisms that reproduce sexually. Due to genetic variation, individuals within a population have slight differences, which might be an advantage if the environment changes. Some individuals might have traits that enable them to survive unusually harsh conditions such as a drought or severe cold. Other individuals might have traits that make them resistant to disease.
Disadvantages	Asexual reproduction results in little genetic variation within a population. Genetic variation can give organisms a better chance of surviving if the environment changes. Another disadvantage of asexual reproduction involves genetic changes, called mutations. If an organism has a harmful mutation in its cells, the mutation will be passed to asexually reproduced offspring. This could affect the offspring's ability to survive.	Sexual reproduction takes time and energy. Organisms have to grow and develop until they are mature enough to produce sex cells. Then the organisms have to form sex cells—either eggs or sperm. Before they can reproduce, organisms usually have to find mates. Searching for a mate can take a long time and requires energy. The search for a mate might also expose individuals to predators, diseases, or harsh environmental conditions.

THREE-DIMENSIONAL THINKING

A fatal disease is spreading through an aquarium containing both fish, which reproduce sexually, and sponges, which reproduce asexually. The disease has been identified in both species. **Construct an explanation** that states which species would be more likely to survive. Write your answer in your Science Notebook.

COLLECT EVIDENCE

What are the advantages and disadvantages of the different types of reproduction, such as that of the sea star at the beginning of the lesson? Record your evidence (C) in the chart at the beginning of the lesson.

A Closer Look: Cloning and the Future

Cloning is a type of asexual reproduction performed in a laboratory that produces identical individuals from a cell or from a cluster of cells taken from a multicellular organism. Because all of a clone's chromosomes come from one parent, the clone is a genetic copy of its parent. Farmers and scientists often use cloning to make copies of organisms or cells that have desirable traits, such as a plant with large flowers. In addition to cloning plants, scientists have been able to clone many animals.

HISTORY › Connection The first mammal cloned from an adult cell was a sheep named Dolly, who was born in 1996. Dolly became a media sensation, sparking the debate around animal cloning. Dolly lived her life at the Roslin Institute in Edinburgh, where she helped researchers understand more about animal cloning. The advances that paved the way for her birth led to the cloning of pigs, deer, and even horses.

Scientists are working to save some endangered species from extinction by cloning. In 2001, the gaur, a wild cattle species seen above, was the first endangered species to be cloned. However, cloning does not add genetic diversity to a species, since a clone is a genetic copy of its parent. Some people are also concerned about the high cost of this technique and ethical issues, such as the possibility of human cloning.

It's Your Turn

Present Research advances in cloning and how cloning is used to save endangered species. Create a presentation for your class that shows the advantages and disadvantages of using cloning to save endangered species. Be sure to include visual displays to strengthen your claims.

Review

Summarize It!

1. **Compare and contrast** asexual reproduction and sexual reproduction by filling in the Venn diagram below.

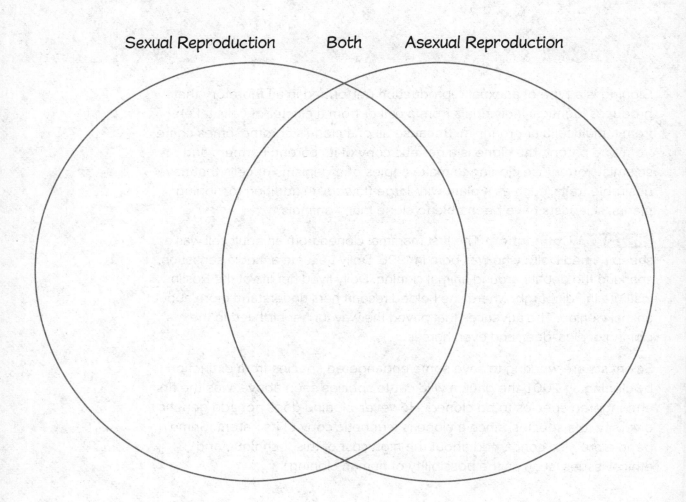

Sexual Reproduction   Both   Asexual Reproduction

2. A tree produces seeds in pods when wind-borne pollen from another tree of the same species reaches the flowers. Each seed contains genetic information so the seed can grow into an adult tree. Which do you predict would be the effect of this process?

    A The tree produces a large number of genetically diverse offspring.

    B The tree produces a large number of genetically identical offspring.

    C The tree produces a small number of offspring that are identical to the female parent.

    D The tree produces a small number of offspring that are identical to the male parent.

Hydras are organisms that live in freshwater environments. They have a tubelike body and a mouth at one end. Around the mouth are stinging tentacles that help to capture food. Depending on the conditions, hydras can reproduce sexually or asexually.

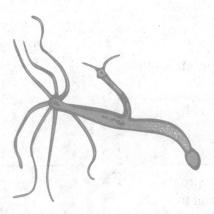

3. Based on your observations, which statement best explains what is happening to the hydra in the figure above?

    A The hydra is reproducing asexually by budding a new hydra.

    B The hydra is reproducing asexually by splitting in two.

    C The hydra is reproducing sexually by grafting to another hydra.

    D The hydra is reproducing sexually by releasing sex cells into the water.

Real-World Connection

4. **Critique** Your friend has, in her aquarium, a single pet Marmorkreb—a type of crayfish. Her pet laid eggs which developed into mature offspring. Your friend thinks that all organisms require two parents for successful reproduction. Critique your friend's argument.

_____

_____

_____

_____

5. **Explain** Recall the three principles of the cell theory. Explain where new cells come from when organisms reproduce, either sexually or asexually.

_____

_____

_____

_____

 Still have questions?
Go online to check your understanding about the different types of reproduction.

 **REVISIT** PAGE KEELEY SCIENCE PROBES
Do you still agree with the person you chose at the beginning of the lesson? Return to the Science Probe at the beginning of the lesson. Explain why you agree or disagree with that person now.

KEEP PLANNING
STEM Module Project Science Challenge

Now that you've learned about sexual and asexual reproduction, go to your Module Project to continue planning your game. You will want to model the differences in asexual and sexual reproduction, as with the Kokanee salmon.

 **EXPLAIN** THE PHENOMENON
Revisit your claim about how the sea star numbers increased. Review the evidence you collected. Explain how your evidence supports your claim.

PERFORMANCE EXPECTATION

LESSON 3 LAUNCH

Growing Up

When animals are born, they don't all grow up to be the same. What determines how a baby animal will grow? Circle the answer that best matches your thinking.

A. Factors in the environment determine how a baby animal will grow.

B. Factors passed from the parents determine how a baby animal will grow.

C. Both factors in the environment and factors passed from the parents determine how a baby animal will grow.

D. There are no factors that determine how a baby animal will grow.

Explain your thinking. Describe your ideas below about factors that determine how animals grow.

_____

_____

_____

_____

_____

_____

You will revisit your response to the Science Probe at the end of the lesson.

Reproduction and Growth of Animals

ENCOUNTER
THE PHENOMENON

What strategies enable these birds of paradise to reproduce successfully, and what affects how they grow?

Bird Dance

▶ **GO ONLINE**

Watch the video *Bird Dance* to see this phenomenon in action.

Observe the birds in the video. What do you think the birds are doing? Brainstorm your ideas below.

EXPLAIN
THE PHENOMENON

Did you see how the birds in the video were dancing to attract mates? Use your observations about the phenomenon to make a claim about reproductive strategies that animals use and factors that affect the growth of their young.

CLAIM

Animals reproduce and ensure the growth of their young by...

 COLLECT EVIDENCE as you work through the lesson.
Then return to these pages to record your evidence.

EVIDENCE

A. What evidence have you discovered to explain how animals, such as the bird of paradise, find mates?

B. What evidence have you discovered to explain how young animals are protected?

MORE EVIDENCE

C. What evidence have you discovered to explain factors that affect how animals grow?

When you are finished with the lesson, review your evidence. If necessary, based on the evidence, revise your claim.

REVISED CLAIM

Animals reproduce and ensure the growth of their young by...

Finally, explain your reasoning for how and why your evidence supports your claim.

REASONING

The evidence I collected supports my claim because...

How do animals find mates?

Most animals are selective about who they choose as a mate. They want to pass certain traits on to their offspring. What are some strategies animals use to attract and select a mate?

These elegant terns are offering gifts of fish to potential mates.

INVESTIGATION

Animal Attraction

Animals attract mates in a variety of ways. Depending on the type of animal, these mating rituals can look very different. In what ways do animals "show off" to attract mates?

GO ONLINE Watch the *Finding the Right One* videos.

Record your observations about how the animals in the videos try to attract mates. What behaviors do you observe?

Want more information?
Go online to read more about the reproduction and growth of animals.

FOLDABLES
Go to the Foldables® library to make a Foldable® that will help you take notes while reading this lesson.

Courtship Behaviors In the videos you observed animals engage in different behaviors in order to find a mate. A **behavior** is the way an organism reacts to other organisms or to its environment. The behaviors you observed in the videos are a form of communication. Animals attract mates by communicating in a variety of ways, including the use of sound, light, chemicals, and body language. Often times animals compete with others of the same species for a mate. Sometimes animals even bring gifts to each other in hopes of getting the attention of one another. What are some examples of these courtship behaviors? Let's investigate!

These sandhill cranes are performing a courtship dance.

INVESTIGATION

Show Off

In order to attract a mate, animals (most often males), frequently "show off" for the other sex. In this activity, your team will investigate an example of an animal reproductive strategy, focusing on external stimuli, sensory receptors, behavioral response, and memory.

1. With your team, choose one of the examples of an animal reproductive strategy provided by your teacher that you would like to investigate, or research and choose one of your own, and record it below.

_____

_____

_____

2. **READING Connection** Use the spaces below to take notes as you conduct your research. Make sure that you use multiple print and digital sources and cite them in your Science Notebook using the format provided by your teacher. For each of the sources used, explain how you assessed its credibility, accuracy, and any possible bias, and cite specific evidence to show that the claims in the sources are supported.

Stimuli:

Sensory receptors:

Male sea lions put on displays to establish and defend their territories for breeding.

Behavioral response:

Memory:

3. **WRITING Connection** Next, prepare a script for a mock debate. Use evidence from your research and scientific reasoning to create a written argument in your Science Notebook that supports an explanation for how the behavior investigated by your team increases the probability of successful reproduction. Make sure you quote or paraphrase in your script to avoid plagiarism. You will present your argument to the class.

4. As the other teams give their presentations, take notes on what you learn.

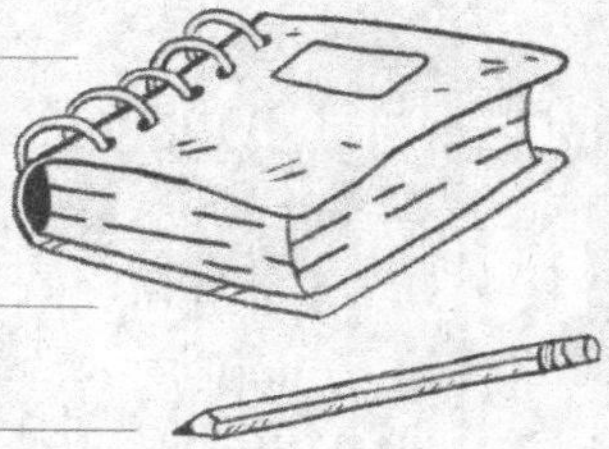

_____

_____

_____

_____

_____

_____

_____

_____

_____

5. After viewing all the teams' presentations, reflect on your team's presentation and report. What do you think you did well? What would you do differently next time?

_____

_____

_____

_____

_____

_____

_____

_____

_____

THREE-DIMENSIONAL THINKING

Explain the cause and effect relationship between courtship behaviors and successful reproduction in animals.

_____

_____

_____

The Spider
Mating Dance

Meet Norman Platnick, a scientist studying spiders.

Norman Platnick is fascinated by all spider species—from the dwarf tarantula-like spiders of Panama to the blind spiders of New Zealand. These are just two of the over 1,800 species he's discovered worldwide.

How does Platnick identify new species? One way is the pedipalps. Every spider has two pedipalps, but they vary in shape and size among the over 46,000 species. Pedipalps look like legs but function more like antennae and mouthparts. Male spiders use their pedipalps to aid in reproduction.

Getting Ready When a male spider is ready to mate, he places a drop of sperm onto a sheet of silk he constructs. Then he dips his pedipalps into the drop to draw up the sperm.

Finding a Mate The male finds a female of the same species by touch or by sensing certain chemicals she releases.

Courting and Mating Males of some species court a female with a special dance. For other species, a male might present a female with a gift, such as a fly wrapped in silk. During mating, the male uses his pedipalps to transfer sperm to the female.

What happens to the male after mating? That depends on the species. Some are eaten by the female, while others move on to find new mates.

▲ Spiders reproduce sexually, so each offspring has a unique combination of genes from its parents. Over many generations, this genetic variation has led to the incredible diversity of spiders in the world today.

◄ Norman Platnick is an arachnologist (uh rak NAH luh just) at the American Museum of Natural History. Arachnologists are scientists who study spiders.

It's Your Turn

Research Select a species of spider and research its mating rituals. What does a male do to court a female? What is the role of the female? What happens to the spiderlings after they hatch? Use images to illustrate a report on your research.

Most animals have specialized behaviors that help them find and attract mates. Some examples of courtship behaviors are in the table below.

They often compete with members of the same species, either physically, or through displays, for a mate. Male mule deer, such as those shown to the right, become very aggressive towards each other when competing for a mate. Peacocks compete by showing off their feathers, and male frigate birds inflate their throat sacks when competing for mates.

Some animals, such as the white female gypsy moth shown to the right, release chemical substances called pheromones that attract males. Many moths and butterflies can detect these pheromones using their sense of smell from up to six miles away.

Other animals, such as birds and the frog shown to the right, use mating songs that gain the attention of mates. Some male birds bring the female a gift of food, such as a male tern bringing a fish to a female. Male fiddler crabs wave their enlarged claws and skitter across the ocean floor in the hopes of getting a female fiddler crab's attention. Male bowerbirds build elaborate nests using brightly colored objects during courtship.

COLLECT EVIDENCE

How do animals, such as the birds of paradise, find mates? Record your evidence (A) in the chart at the beginning of the lesson.

How are young animals protected?

You've learned about courtship behaviors that increase the probability of finding a mate, but what are behaviors that increase the chance that offspring will survive after animals reproduce?

A Western gull and her chicks

INVESTIGATION

Staying Safe

When goslings, or baby geese, see a bird in the air that has a different wingspan or shape than the parent goose, they duck down.

1. Look at the images of the three birds in flight. Describe the differences between each silhouette.

_____

_____

_____

_____

2. Choose at least two characteristics that are different for each bird.

_____

_____

_____

_____

3. How could recognizing differences help a gosling survive?

_____

_____

_____

INNATE v. LEARNED
BEHAVIORS

Parents and offspring both engage in certain behaviors that increase the probability that young animals will survive. Some are inherited and some are learned.

INNATE BEHAVIOR
is a behavior that is inherited rather then learned.

LEARNED BEHAVIOR
is a behavior that develops through experience or practice.

SPIDERS
Spiders instinctively know how to build webs in order to catch food.

BIRDS
Birds learn how to fly through trial and error and reinforcement from their parents.

TADPOLES
When tadpoles hatch, they already know how to swim. They can avoid danger as soon as they are born.

TURTLES
Female sea turtles return to the beach where they were born to lay their eggs. These turtles imprinted on the beach.

In this lab, you will work as a team to complete a challenge—building your own birds' nest! Will your nest be successful in protecting eggs and young birds?

Safety

Materials

plastic grocery bag	cardboard (at least 24" × 24")
small plastic baggie	containers for water
plastic spoon	bird nest building materials
plastic tarp	eggs

Procedure

1. Read and complete a lab safety form.

2. As a class, identify the criteria and constraints of the challenge. Record your responses in the boxes below.

Criteria:

Constraints:

  Time:

  Materials:

3. **READING Connection** As a team, research different types of birds' nests and the ways in which different species of birds build their nests.

4. Decide which species of birds' nest your team will build.

5. What materials will you need to gather to build your nest?

Procedure, continued

6. Plan how you will build your nest. Draw a diagram below, labeling the parts, and including the location where the nest will be built.

7. Gather your materials.

8. Build your nest.

9. Test your nest by lifting it up to see if it holds together. Put eggs in the nest and try this test again. Would the eggs fit in the nest with the parent birds? Gently shake the nest to see if the eggs would be safe and stable inside the nest. Use your hands to feel inside the nest. Will the chicks be comfortable in your nest? Record your observations in the Data and Observations section.

10. Follow your teacher's instructions for proper cleanup.

Data and Observations

Analyze and Conclude

11. How did your nest meet the criteria and constraints of the project? What improvements could you make to your nest?

_____

_____

_____

_____

_____

_____

_____

12. After observing the other teams' nests, reflect on the challenge. Did your nest look like an actual bird's nest? Which group do you think had the best nest, and why? What do you think your group did well? What could you improve on?

_____

_____

_____

_____

_____

_____

_____

_____

_____

Protecting Offspring Birds aren't the only animals that engage in nest building behaviors in order to protect their young. Some mammals, amphibians, fish, reptiles, and insects have the instinct to protect their eggs and their young, or even themselves, by building nests. Animals build different types of nests, or dens, out of a variety of materials for protection from predators and the environment.

Another example of a nurturing behavior is herding. Many animals herd their offspring to ensure the young animals are not left behind and are safe from predators. Elephants take turns watching over each other's babies so the mother can take breaks!

Some animals, such as musk oxen, circle their young with their horns facing out to protect them from predators. Bison form two circles around their young for protection—the females form a circle around the young, and males form a circle around the females.

A wild horse protecting its young

COLLECT EVIDENCE

How are young animals, such as bird of paradise chicks, protected? Record your evidence (B) in the chart at the beginning of the lesson.

What factors affect how baby animals grow?

When baby birds hatch, will they all grow up to be the same? No, there are many factors that determine how animals will grow. Some of these factors are genetic—traits are inherited from the parents, and some factors are environmental—the animal's environment influences how it will grow. Can some factors be both genetic and environmental?

INVESTIGATION

Just Grow with It

In this activity, you and your team will explain how genetic and environmental factors influence the growth of an animal.

1. Choose an animal that you are somewhat familiar with (does anyone on your team have any pets?) to base your explanation on.

2. Next, brainstorm different traits of that animal.

3. With your team, research the traits brainstormed above, and use the space below to place each trait on the spectrum of genetic or environmental. Can some traits be influenced by both genetic and environmental factors? If a trait is influenced more by genetics, place it towards the top of the arrow, and if it is influenced more by the environment, place it more towards the bottom of the arrow.

Genetic

Environmental

4. Complete the graphic organizer below to determine causes and effects of the growth of your animal.

Environmental Causes	Genetic Causes

Mechanisms (What processes lead to the effects?)	Mechanisms (What processes lead to the effects?)

Effects	Effects

5. Write a short report that explains how genetic and environmental factors influence the growth of your animal. Your explanation should be based on valid and reliable evidence from your sources.

_____

_____

_____

_____

_____

_____

_____

_____

_____

_____

Factors Influencing the Growth of Animals Both genetic and environmental factors affect the growth of animals. Genetic factors that affect growth are inherited from the parents. They are determined by genes. For example, the type of fur a dog has (straight, curly, wiry) is determined by the genes of the parents. Some traits, such as obedience of a dog, are influenced by the environment. Some traits of animals are influenced by both genetic and environmental factors. The weight of an animal is partially determined by genes, and partially determined by its diet. For many traits, it is still unknown whether they are determined by genetics, the environment, or both. Since many factors influence the growth of animals, the traits of animals can often not be predicted.

COLLECT EVIDENCE
What factors affect the growth of animals, such as birds of paradise? Record your evidence (C) in the chart at the beginning of the lesson.

Review

Summarize It!

1. **Organize** what you have learned about animal behaviors that increase the probability of successful reproduction and different factors that affect the growth of young animals.

Behaviors That Increase the Probability of Successful Reproduction

Courtship	Protecting Young

Factors That Affect the Growth of Young Animals

Environmental Factors	Genetic Factors

Three-Dimensional Thinking

In order to attract a mate, male peacocks fan out their colorful feathers and dance. Females tend to choose males that have larger displays of feathers and feathers with more eyespots. The peahen then builds her nest by scraping a hole in the ground in a hidden area. Once the chicks hatch, the peahen stays close to them, teaching them what foods to eat and defending them from predators.

2. Which of the following is a courtship behavior that increases the probability of successful reproduction for the peacock?

   A   fanning feathers

   B   nest building

   C   protecting from predators

   D   all of the above

Observe the hamsters' environment below.

3. Which of the following is NOT an environmental factor that would affect the hamsters' growth?

   A   the amount of food the hamster is given

   B   gene for fur color

   C   the amount of time spent on the exercise wheel

   D   interactions with other hamsters

Real-World Connection

4. Argue Your friend thinks that successful reproduction of animals depends only on the behaviors of the parents. Refute his claim using evidence and reasoning and present your argument orally to the class.

_____

_____

_____

_____

5. Explain how farmers can benefit by learning about factors that affect the growth of animals.

_____

_____

_____

_____

 Still have questions?
Go online to check your understanding about the reproduction and growth of animals.

REVISIT

 **SCIENCE PROBES**

Do you still agree with the answer you chose at the beginning of the lesson? Return to the Science Probe at the beginning of the lesson. Explain why you agree or disagree with that answer now.

KEEP PLANNING

STEM Module Project
Science Challenge

Now that you've learned about animal reproduction and growth, go back to your Module Project to continue planning your game. Include behaviors that affect the probability of successful reproduction, such as the salmon returning to their home stream to spawn.

EXPLAIN
THE PHENOMENON

Revisit your claim about how animals reproduce and grow. Review the evidence you collected. Explain how your evidence supports your claim.

What does a flower do?

Mrs. Gonzales has a flower garden. Her neighbors always remark how beautiful her flowers are. One day Mrs. Gonzales asked her neighbors, "Why do plants produce flowers? What is their main purpose?" They each had a different idea. This is what they said:

Mrs. Flores: I think plants produce flowers in order to attract bees.

Mr. Myer: I think plants produce flowers so they can make their own food.

Ms. Bricker: I think plants produce flowers so they will look pretty.

Mr. Pappas: I think plants produce flowers for reproduction.

Mr. Frist: I think plants produce flowers so they can respond to their environment.

Ms. Chai: I think plants produce flowers so people can use them for special occasions.

Mrs. Meehan: I think plants produce flowers so that other plants can detect them by smell.

Which person do you agree with the most? _____
Explain your thinking about the function of a plant's flower.

_____

_____

_____

_____

_____

_____

You will revisit your response to the Science Probe at the end of the lesson.

Reproduction and Growth of Plants

ENCOUNTER
THE PHENOMENON
What structures enable this purple tansy plant to successfully reproduce, and what affects how it grows?

Observe the parts of the plant provided by your teacher. What do you think are the purposes of these structures? Record your thoughts below.

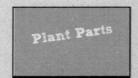

GO ONLINE
Watch the video *Plant Parts* to see this phenomenon in action.

EXPLAIN
THE PHENOMENON

Did you see how the plant has many different structures? Use your observations about the phenomenon to make a claim about how plants reproduce and factors that affect how they grow.

CLAIM

Plants reproduce and grow by...

 COLLECT EVIDENCE as you work through the lesson. Then return to these pages to record your evidence.

EVIDENCE

A. What evidence have you discovered to explain how plants, such as the purple tansy, reproduce?

B. What evidence have you discovered to explain how plants, such as the purple tansy, find mates and spread seeds?

MORE EVIDENCE

C. What evidence have you discovered to explain factors that affect how plants, such as the purple tansy, grow?

When you are finished with the lesson, review your evidence. If necessary, based on the evidence, revise your claim.

REVISED CLAIM

Plants reproduce and grow by...

Finally, explain your reasoning for how and why your evidence supports your claim.

REASONING

The evidence I collected supports my claim because...

How do different types of plants reproduce?

In the Encounter the Phenomenon activity, you observed plant structures. Some of those structures aid in plant reproduction. Do all plants have those same structures? Do all plants reproduce in the same way? Let's explore.

 Want more information?
Go online to read more about the reproduction and growth of plants.

FOLDABLES
Go to the Foldables® library to make a Foldable® that will help you take notes while reading this lesson.

LAB Seeds of Thought

Safety

Materials

pine cone

apple

knife

Procedure

1. Read and complete a lab safety form.

2. Examine a pine cone and find the seeds. Draw a sketch of a pine cone seed, and record the location of the seeds in the Data and Observations section on the next page.

3. Examine an apple. Using a knife, carefully cut the apple in half to locate the seeds. Draw a sketch of an apple seed, and record the location of the seeds in the Data and Observations section on the next page.

4. Follow your teacher's instructions for proper cleanup.

Data and Observations

Analyze and Conclude

5. Compare and contrast the seeds in the pine cone and in the apple.

_____

_____

_____

_____

_____

6. Thinking about the structures of both of the types of seeds, infer how both of the plants might reproduce.

_____

_____

_____

_____

Types of Plant Reproduction Plants can reproduce either asexually, sexually, or both ways. Asexual reproduction occurs when a portion of a plant develops into a separate new plant. Sexual reproduction occurs when a plant's male reproductive cell (sperm) combines with a plant's female reproductive cell (egg). The way a plant reproduces depends on the structures it has.

These hens and chicks can reproduce without seeds, or asexually. New "chicks" can grow from the stolons on the main "hen" plant.

Seedless Plants Not all plants grow from seeds. The first land plants to inhabit Earth most likely were seedless plants—plants that grow from spores, not from seeds. Mosses and ferns are examples of seedless plants found on Earth today.

Seed Plants There are two groups of seed plants—flowerless seed plants and flowering seed plants. Both produce seeds that result from sexual reproduction. The plants produce pollen grains, which contain sperm. They also produce female structures, which contain one or more eggs. **Pollination** occurs when pollen grains land on a female plant structure of the same species. If the pollen joins with an egg, fertilization occurs and a seed develops. In nonflowering plants, the pollen is produced by the male cone, and the eggs are contained within the female cone. In flowering plants, the female reproductive organ is the pistil, and the male reproductive organ is the stamen.

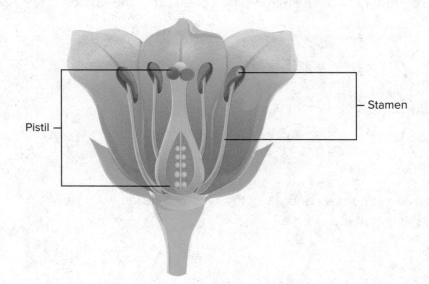

Pistil —

— Stamen

COLLECT EVIDENCE

How do plants, such as the purple tansy, reproduce? Record your evidence (A) in the chart at the beginning of the lesson.

How can plants find mates and spread seeds if they cannot move?

Plants can't walk around to find mates or spread seeds, so how do plants reproduce successfully? And why aren't all plant offspring right next to the parent plant? There are a variety of different ways pollination can occur and seeds can spread.

ENGINEERING

LAB Blowing in the Wind

In order for reproduction to be successful, seeds must be dispersed to places where resources, such as light, food, water, and space, are available. In this lab, you will work with a partner to design a seed structure that can by carried by wind as far as possible.

Safety

Materials

8 ½" × 11" paper

scissors

tape

lima bean

fan

meterstick

Procedure

1. Read and complete a lab safety form.

2. As a class, identify the criteria and constraints of the challenge.

Criteria:

Constraints:

Time:

Materials:

Procedure, continued

3. Plan your seed structure. Draw a diagram of the structures and label the components.

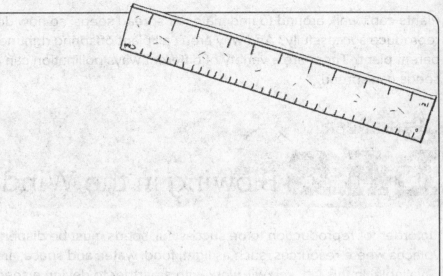

4. Build your seed structure.

5. Test your design by dropping it in front of a fan three times and measuring the distance it travels. Then find the average. Record your data in the Data and Observations section on the next page.

6. What can you do to improve your design? Make any necessary changes and record them below.

7. Test your improved design three times and find the average distance the seed structure travels. Record your data in the Data and Observations section on the next page.

8. Follow your teacher's instructions for proper cleanup.

Data and Observations

	Trial 1	Trial 2	Trial 3	Average
Design 1				
Design 2				

Analyze and Conclude

Record your responses in your Science Notebook.

9. How was this activity similar to and different from what might occur in nature?

10. **WRITING ▸ Connection** Construct an argument using evidence and reasoning, to support the idea that plant structures aid in reproduction.

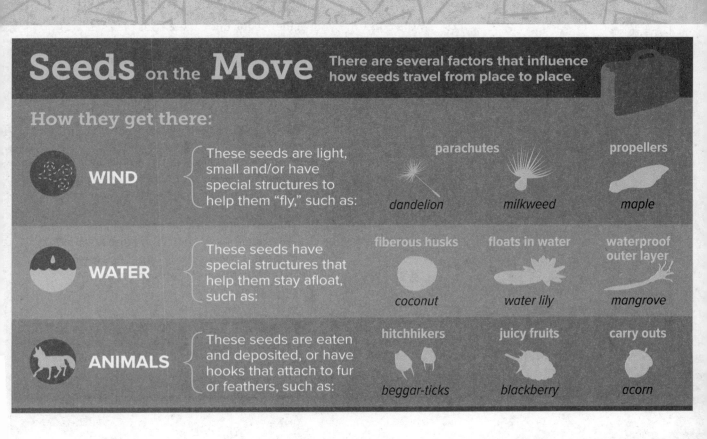

Seeds on the Move

There are several factors that influence how seeds travel from place to place.

How they get there:

WIND — These seeds are light, small and/or have special structures to help them "fly," such as:
- parachutes — *dandelion*
- *milkweed*
- propellers — *maple*

WATER — These seeds have special structures that help them stay afloat, such as:
- fiberous husks — *coconut*
- floats in water — *water lily*
- waterproof outer layer — *mangrove*

ANIMALS — These seeds are eaten and deposited, or have hooks that attach to fur or feathers, such as:
- hitchhikers — *beggar-ticks*
- juicy fruits — *blackberry*
- carry outs — *acorn*

Read a Scientific Text

Bees play an important role in pollination. As they move from flower to flower collecting nectar for food, they transfer pollen, enabling the plants to reproduce. How is climate change affecting this relationship between bees and pollination?

CLOSE READING

Inspect

Read the passage *Buzzing About Climate Change.*

Find Evidence

Reread the passage. Underline the relationships between events that can be described as a cause and effect relationship. Looking at the evidence you underlined, determine the main idea of the passage. Write a summary of the passage in your Science Notebook.

Make Connections

Communicate With a partner, discuss how changes in Earth's temperature affect pollination. Have you heard any other discussion of this phenomenon?

PRIMARY SOURCE

Buzzing about Climate Change

[...] Nature could tell us thousands of stories about how climate change is affecting life on Earth. [...] Several years ago, NASA oceanographer and amateur beekeeper Wayne Esaias realized he was overhearing one of those stories. The talk of climate change was coming from his bees. [...] Esaias believes that a beehive's seasonal cycle of weight gain and loss is a sensitive indicator of the impact of climate change on flowering plants. [...] The most important event in the life of flowering plants and their pollinators—flowering itself—is happening much earlier in the year than it used to. [...]

[...] Agriculture depends on managed honeybees not only because [...] our industrial-scale system of crop production hinges on huge numbers of pollinators being available in a very limited window of time [...]

"Flowering plants and pollinators co-evolved. [...] Some species of pollinators have co-evolved with one species of plant, and the two species time their cycles to coincide [...]"

The concern is that in thousands upon thousands of cases, we don't really know what environmental and genetic cues plants and pollinators use to manage this synchrony. [...] some plant-pollinator pairs in a particular area likely do respond to the same environmental cues, and it's reasonable to expect they will react similarly to climate change. But other pairs use different cues, the pollinator emerging in response to air temperature, for example, while the plant flowers in response to snow melt. [...] There is no guarantee that the thousands of plant-pollinator interactions that sustain the productivity of our crops and natural ecosystems won't be disrupted by climate change.

[...] Between urbanization and global warming from greenhouse gases, temperatures will continue to rise in coming years; the acceleration in flowering times that Esaias' honeybees have documented so far may not be the end of the changes.

Source: NASA

COLLECT EVIDENCE

How do plants, such as the purple tansy, find mates and spread seeds? Record your evidence (B) in the chart at the beginning of the lesson.

What factors affect how plants can grow?

You've learned about the different ways in which plants reproduce. What are different factors that affect how offspring plants grow?

INVESTIGATION

Testing Plant Growth

In this investigation, you will observe and collect data on plants in different environments. How does high salinity, cold, heat, or drought affect the growth of plants?

1. Inspect the plants that your teacher has prepared for this investigation, and record your observations about each plant in the chart.

Treatment	Plant height	Number of leaves	Wilting? Yes/No	Color of leaves	Root length
Control					
Drought					
Cold					
Saline					
Heat					

2. After analyzing and interpreting your data, construct an explanation in your Science Notebook on different factors that affect the growth of plants. Use evidence from the investigation to support your answer.

THREE-DIMENSIONAL THINKING

Analyze and interpret the data from the Investigation *Testing Plant Growth* to explain the cause and effect relationship between environmental factors and plant growth. Record your response in your Science Notebook.

In the investigation *Testing Plant Growth* you observed environmental factors that affect how plants grow. A tropism (TROH pih zum) is a response that results in plant growth toward or away from a stimulus. The growth of a plant toward or away from light is called **phototropism.** A plant has a light-sensing chemical that helps it detect light. The response of a plant to touch is called **thigmotropism** (thihg MAH truh pih zum). The vine growing up the fence in the photo clings to the fence in response to touching it. The response of a plant to gravity is called **gravitropism.** Stems grow away from the pull of gravity, while roots grow toward gravity.

Genetic factors also affect how plants grow. Plants inherit genes from their parents that determine traits, such as what color flowers they will have and where on the plant the flowers will bloom.

▶ **GO ONLINE** for additional opportunities to explore!

Investigate factors that affect plant growth by performing one of the following activities.

☐ **Observe** phototropism and gravitropism in action in the **Lab** *How does an external stimulus affect the growth of a plant?*

OR

☐ **Determine** how light affects the growth of plants in the **Lab** *How important is light to the growth of plants?*

COLLECT EVIDENCE
What factors affect how plants, such as the purple tansy, grow? Record your evidence (C) in the chart at the beginning of the lesson.

A Closer Look: Growing Plants in Space

Have you ever seen a greenhouse, such as the one shown in the photo to the right? Greenhouses are used for growing plants in cool climates. They are made out of material that allows sunlight to enter but does not allow heat to escape. The transparent material reflects energy back into the greenhouse and stops winds from carrying it away.

Could greenhouses be used to grow plants on the Moon? The conditions in space are much different than here on Earth. Plants don't have the resources in space, such as light, moisture, and the right temperature, that they need to survive. Scientists have created structures called plant growth chambers in order to be able to grow plants in space to be used as food. These growth chambers create an environment similar to the environment on Earth.

It's Your Turn

Design Imagine that you are an engineer tasked with creating a plant growth chamber to be used to grow plants in space. What materials would you use? Draw a plan and label the parts of your plant growth chamber.

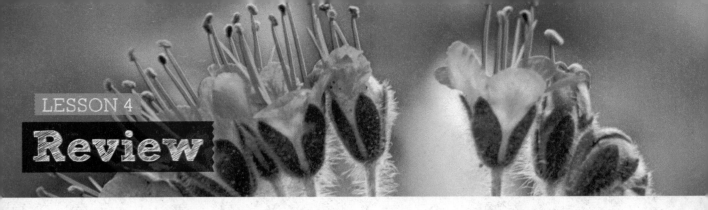

Review

Summarize It!

1. **Create** a graphic organizer below that shows how specialized structures increase the probability of successful reproduction in plants, and factors that affect their growth.

Three-Dimensional Thinking

2. Which of the following is a plant structure that increases the probability of successful reproduction?

A

B

C

D

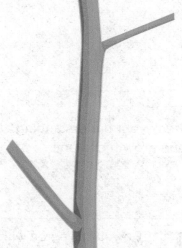

Mr. Blake is preparing to plant his yearly corn crop. In order to produce the best crop yield possible, he considers a variety of factors that can affect the growth of the corn.

3. Which of the following is a factor that can affect the growth of the corn crops?

A gene for color of kernels

B amount of water given

C the space available for the plants to grow

D all of the above

Real-World Connection

4. **Argue** Honeybee populations have been declining rapidly in the past decade. Construct an argument that this can affect food production.

_____

_____

_____

_____

_____

5. **Explain** when it could be beneficial to understand genetic factors that affect the growth of plants.

_____

_____

_____

_____

 Still have questions?
Go online to check your understanding about the reproduction and growth of plants.

REVISIT  **PAGE KEELEY SCIENCE PROBES**

Do you still agree with the neighbor you chose at the beginning of the lesson? Return to the Science Probe at the beginning of the lesson. Explain why you agree or disagree with that neighbor now.

EXPLAIN THE PHENOMENON

Revisit your claim about how plants reproduce and grow. Review the evidence you collected. Explain how your evidence supports your claim.

PLAN AND CREATE

STEM Module Project
Science Challenge

Now that you've learned about plant reproduction and growth, go back to your Module Project to continue planning your game. Your goal is to include factors in the game that affect the growth of plants and animals, such the salmon at the beginning of the module.

PERFORMANCE EXPECTATION

Get Your Game Face On

A scientific toy company has asked your team to develop and test a game that models the differences in offspring produced through asexual and sexual reproduction. The goals of the game are for players to ensure the successful reproduction of an alien species that can reproduce through different methods.

Your game will include behaviors as well as specialized structures that affect the probability of reproduction, environmental and genetic factors influencing the growth of organisms, and senses that help with reproduction and survival, such as those that help Kokanee salmon return to their home streams to spawn.

Planning After Lesson 1

What kind of game will you create to model reproduction and factors that affect an organism's success of reproduction?

Planning After Lesson 1, continued

Answer the following questions to help guide the development of your game:

How will you incorporate parent genotypes in your game?

How many traits will you provide information about?

How will players know which trait is dominant and which is recessive?

Planning After Lesson 2

Explain how your game will incorporate the differences in offspring
produced through asexual and sexual reproduction.

Planning After Lesson 3

Special behaviors are important for reproductive success in animals.
Assume that similar behaviors affect reproduction and survival of offspring
in your alien species. Make *Behavior* cards that will be used in your game in
some way, and record them below.

In your Science Notebook, gather information about how the senses aid in
reproduction. How will you make this concept a playable part of your game?

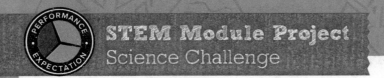
Planning After Lesson 4

Environmental factors influence the growth of plants and animals. Assume that similar factors affect your alien species. Make *Environmental Factors* cards to include with your game and record them below.

Specialized structures are important for reproductive success in plants. Assume that your alien species also has some structures that aid in reproduction. Make *Structures* cards describing some of these structures and record them below.

Develop Your Game

Construct your game. Once your game is complete, test it out with your classmates. Then, identify the model elements in the table below.

Model Elements	Descriptions
Components (What are the different parts of my model?)	
Relationships (How do the components of my model interact?)	
Connections (How does my model help me understand the phenomenon?)	

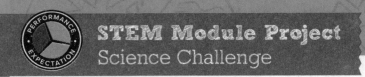
Create Your Presentation

Analyze and evaluate your game before you present to the toy company.
How do the relationships and connections you identified in your game lead
to a better understanding of how animals reproduce?

_____

_____

_____

_____

_____

At any point during the development of your game, did you need to return
to previous plans and make changes in light of new information?

_____

_____

_____

_____

_____

_____

How could your game help explain the photo of the puppy and its parent at
the beginning of the module project?

_____

_____

_____

_____

*Congratulations! You've
completed the Science
Challenge requirements!*

Module Wrap-Up

REVISIT
THE PHENOMENON

Using the concepts you have learned throughout this module, explain how Kokanee salmon reproduce and grow.

OPEN INQUIRY

What are one or two questions you still have about the phenomenon?

_____

_____

_____

_____

Choose the question that interests you the most. Plan and conduct an investigation to answer this question.

Glossary

Multilingual Glossary

A science multilingual glossary is available on the science website. The glossary includes the following languages.

Arabic	Hmong	Tagalog
Bengali	Korean	Urdu
Chinese	Portuguese	Vietnamese
English	Russian	
Haitian Creole	Spanish	

Cómo usar el glosario en español:
1. Busca el término en inglés que desees encontrar.
2. El término en español, junto con la definición, se encuentran en la columna de la derecha.

Pronunciation Key

Use the following key to help you sound out words in the glossary.

a . . . . back (BAK)	Ew. . . . . food (FEWD)
ay . . . . day (DAY)	yoo . . . . pure (PYOOR)
ah . . . . father (FAH thur)	yew. . . . . few (FYEW)
ow. . . . . flower (FLOW ur)	uh . . . . comma (CAH muh)
ar . . . . car (CAR)	u (+ con). . . . . . . rub (RUB)
E . . . . less (LES)	sh . . . . shelf (SHELF)
ee . . . . leaf (LEEF)	ch . . . . nature (NAY chur)
ih . . . . trip (TRIHP)	g . . . . gift (GIHFT)
i (i + com + e) . . . . idea (i DEE uh)	J . . . . gem (JEM)
oh. . . . . go (GOH)	ing . . . . sing (SING)
aw. . . . . soft (SAWFT)	zh . . . . vision (VIH zhun)
or . . . . orbit (OR buht)	k . . . . cake (KAYK)
oy . . . . coin (COYN)	s . . . . seed, cent (SEED)
oo . . . . foot (FOOT)	z . . . . zone, raise (ZOHN)

English — A — Español

allele/gene   **alelo/gen**

allele (uh LEEL): a different form of a gene.

asexual reproduction: a type of reproduction in which only one parent organism or part of that organism produces a new organism.

alelo: forma diferente de un gen.

reproducción asexual: tipo de reproducción en el que a partir de un organismo parental o parte de éste, se produce un organismo nuevo.

B

behavior: the way an organism reacts to other organisms or to its environment.

budding: a form of asexual reproduction where an organism grows on the body of its parent.

comportamiento: forma en la que un organismo reacciona hacia otros organismos o hacia su medioambiente.

gemación: forma de reproducción asexual en la cual un organismo crece en el cuerpo del organismo parental.

D

dominant (DAH muh nunt) **trait:** a genetic factor that blocks another genetic factor.

rasgo dominante: factor genético que bloquea otro factor genético.

G

gene (JEEN): a section of DNA on a chromosome that has genetic information for one trait.

gen: parte del ADN en un cromosoma que contiene información genética para un rasgo.

genetics: the study of how traits are passed from parents to offspring.

genotype (JEE nuh tipe): the alleles of all the genes on an organism's chromosomes; controls an organism's phenotype.

gravitropism: the response of a plant to gravity.

genética: estudio de cómo los rasgos pasan de los padres a los hijos.

genotipo: de los alelos de todos los genes en los cromosomas de un organismo, los controles de fenotipo de un organismo.

gravitropismo: respuesta de las plantas a la gravedad.

H

heterozygous (he tuh roh ZI gus): a genotype in which the two alleles of a gene are different.

homozygous (hoh muh ZI gus): a genotype in which the two alleles of a gene are the same.

heterocigoto: genotipo en el cual los dos alelos de un gen son diferentes.

homocigoto: genotipo en el cual los dos alelos de un gen son iguales.

P

pedigree: a model that shows genetic traits that were inherited by members of a family.

phenotype (FEE nuh tipe): how a trait appears or is expressed.

phototropism: the growth of a plant toward or away from light.

pollination (pah luh NAY shun): the process that occurs when pollen grains land on a female reproductive structure of a plant that is the same species as the pollen grains.

pedigrí: modelo que muestra los rasgos genéticos heredados por los miembros de una familia.

fenotipo: forma como aparece o se expresa un rasgo.

fototropismo: crecimiento de una planta hacia o lejos de una luz.

polinización: proceso que ocurre cuando los granos de polen posan sobre una estructura reproductiva femenina de una planta que es de la misma especie que los granos de polen.

R

recessive (rih SE sihv) trait: a genetic factor that is blocked by the presence of a dominant factor.

regeneration: a type of asexual reproduction that occurs when an offspring grows from a piece of its parent.

rasgo recesivo: factor genético boqueado por la presencia de un factor dominante.

regeneración: tipo de reproducción asexual que ocurre cuando un organismo se origina de una parte de su progenitor.

S

sexual reproduction: type of reproduction where the genetic material from two different cells combine, producing an offspring.

reproducción sexual: tipo de reproducción en el que a partir de un organismo parental o parte de éste, se produce un organismo nuevo.

T

thigmotropism: the response of a plant to touch.

tigmotropismo: respuesta de una planta al toque.

V

vegetative reproduction: a form of asexual reproduction in which offspring grow from a part of a parent plant.

reproducción vegetativa: forma de reproducción asexual en la cual el organismo se origina a partir de una planta parental.

Italic numbers = illustration/photo
Bold numbers = vocabulary term
lab = indicates entry is used in a lab
inv = indicates entry is used in an investigation
smp = indicates entry is used in a STEM Module Project
enc = indicates entry is used in an Encounter the Phenomenon
sc = indicates entry is used in a STEM Career